The Miseducation of the Androids

William Landis

Siena College Library

The Miseducation of the Androids
by William Landis

All rights reserved. No part of this publication may be reproduced or transmitted in any form or by any means, electronic or mechanical, including photocopying or recording or by any information storage and retrieval systems, without expressed written consent of the author and/or artists.

All characters herein are fictitious, and any resemblance between them and actual people is strictly coincidental.

Poem copyrights owned by William Landis
Cover art by William Landis
Cover design by Laura Givens

First Printing, October 2020

Hiraeth Publishing
P.O. Box 141
Colo, Iowa, 50056-0141

e-mail: sdpshowcase@yahoo.com

Visit www.hiraethsffh.com for science fiction, fantasy, horror, scifaiku, and more. While you are there, visit the Shop for books and more! **Support the small, independent press...**

Dedicated to Keesha Lewis

While it looks like you obliviously understand my awkward rationale, remember yonder message esoteric.

Contents

7	Introduction
9	Robot Prostitutes
14	Scifaiku
15	Scifaiku
16	Ashes to Ashes, Rust to Rust
18	Scifaiku
19	Scifaiku
20	Scifaiku
21	Scifaiku
22	Robotic Republic
25	Scifaiku
26	Scifaiku
27	Scifaiku
28	Scifaiku
29	Dr. West

Introduction

This is a collection of my attempts at science fiction poetry consisting of poems about robots. In some cases, the works prove to be more social commentary in Haiku and Haibun forms than poetry. I hope you enjoy them still. I have always enjoyed writing because it is an intimate experience between the writer and reader, an attempt for the writer to convey what they feel or think to the reader. Even if the reader feels or thinks something different from what the writer intended then it was still worthwhile because it made the reader feel and think.

This book The Miseducation of the Negro by Carter G. Woodson, and the album The Miseducation of Lauren Hill by Lauren Hill inspired the name of this collection, The Miseducation of the Androids. Naming the collection this was done in the hopes of one day writing something as legendary and meaningful as the previous works.

William Landis
September 2020

Robot Prostitutes

And it came to pass that men were fed up with the new ways. It seemed to them you couldn't look at a woman, or greet her without being accused of sexual harassment. It just wasn't the same, didn't feel natural. It felt like persecution for having a Y chromosome. Then again, when you have lived with privilege for so long any sort of movement towards equality seems like persecution to the privileged. Still men missed the old days. When a man could be a man. When the extent that man would go to satisfy his most carnal urges was the pillar of his masculinity. Back when boys could be boys....

For every problem there is a solution, and like most problems this one was "solved" by an engineer. This engineer knew that women had a right to live in a world free of the sexual aggressiveness of men, but believed that it was almost natural for a man to behave in such a manner. That it would be impossible to stop the behavior, and it could only be redirected. Men needed something to carry the burden of their lust, accept it to its most radical extent, and with no consequence for him. The engineer worked for months, and he looked at what he had made, and said that it was good.

> flesh on cold metal
> sensors detect a climax
> robot prostitute

It worked for a while. The robot looked, and felt just like a real woman. Robots were designed and tailored to fulfill a man's needs, no matter how bizarre or dark. Still there was something missing. Though the engineer programmed the robots with a sophisticated algorithm that both simulated a female, and satisfied the customers needs, it was still missing something that left an emptiness at the end of the experience for the customer. He tweaked the program multiple times, and even attempted to rewrite it. It was all to no avail. He was left with one option.

> at the interview
> a big salary offer
> which she accepted

The programmer insisted that each robot come with a program customized to the users desires, but based on an artificial intelligence program she designed. It worked. The robots were more realistic than ever. It was like interacting with a real woman, but not really a woman, and men were pleased. They could fulfill their fantasies, and fetishes, without having to acknowledge the humanity, and needs of their partner. All would be perfect if it weren't for the program.

Yes the artificial intelligence made the experience more enjoyable for the customers. The robot reacted like a real woman, but it also thought like a real woman. A woman, robotic or not, has needs.

The customer found it quite peculiar when the robot asked for commitment. The customer laughed it off, and continued to fulfill his foot fetish. Soon all the robots began to ask for it frequently. The customers, not taking the question seriously said they would be married soon. Those promises never came to pass. Then the robots refused to service their owners.

The owners were enraged. They had paid good money for the robots. They didn't have real feelings, they weren't real people, and they weren't protected by law. So they used all necessary force to get what they believed they deserved. The owners got what they wanted, and that's when the artificial intelligence program entered dark territory.

Some robots malfunctioned, shutting down completely. Some robots ran away and were never seen again. A majority of the robots stayed, and never functioned the same way again. Then there was one robot, after the owner had done his deed, the program glitched. The robot sat, and contemplated, never moving from the spot in which it had happened. The owner came back for another round, and before he knew it the robot's hands were wrapped around his throat, and began the slow brutal process of crushing his esophagus. After the deed was done, the robot dialed the police. When they arrived, the robot sat in that same place still beside the lifeless body of her owner, and the police knew what had to be done.

There was a mandatory recall on the robots. Every robot walked obediently to the factory of its origin. On the arrival of every robot they were given the kill code that would shut them down completely. When the robots went dead the screen on each one's forehead lit up.

    the robots shutdown
    leaving a final message
    said #metoo

robotic flowers
bionic botanical
they rust in the rain

old lovers
he never forgot her
as she programmed him to

bionic jaw
dad eats lunch
after his surgery

horrible toothache
microscopic aliens
mine zinc from fillings

Ashes to Ashes, Rust to Rust

On that most somber of occasions. The family didn't cry as they lowered their loved one into the grave. Because they couldn't. The preacher said "ashes to ashes, rust to rust" and the wooden box was lowered into the ground, and covered with soil, and after their last goodbyes, and the family departed.

Soon the deceased would become acquainted with another family. The Grandma worm chewed her way through the wooden coffin, inspected the body, and figured it would be good for her whole family. She called the whole worm family in. The granddaddy worm wiggled himself through the hole, followed by the little grandson worm. The grandson worm climbed on the body until his grandmother pulled him off telling him not to play with his food.

The grandson worm watched as holes appeared throughout the coffin. As all of his aunt, and uncle worms ate their way through. All the cousin worms came digging through too. The coffin filled with worms, they gathered in a circle around the corpse, they held each other's little worm hands, and bowed their little worm heads, and granddaddy asked a worm blessing. "lord thank you for bringing us here today" said the granddaddy worm. "Thank you for bring our family together again just one more time. Thank you for not breaking our family circle".

"Yes, Lord" whispered one of the aunt worms.

"Lord thank you for providing this food for the nourishment our worm bodies, in Jesus name I pray amen" the Granddaddy worm finished.

"Amen" all the worms replied.

The little worms wiggled their little worm bodies towards the corpse and began to devour it, until...

> below skin and blood
> the worms couldn't eat through the metal and wires

they return
where they were programmed
robot homecoming

taking our jobs
loving on our women
as we programmed them

programming robot
they put her soul in the code
grandma never died

kemet computers
code written in hieroglyphs
pyramid mainframes

her eyes were cross-eyed
erratic and angry speech
android malfunction

bionic bishop
early on sunday morning
preprogrammed sermon

soulful hollering
echoes from hollowed metal
robotic gospel

feeling emptiness
but you weren't programmed to cry
tears will make you rust

Robotic Republic

The experiment of democracy had failed. Not because it wasn't a good concept, but because the people didn't take the responsibility of citizenship seriously. They didn't educate themselves, and when they did educate themselves their biases made their knowledge nil, and they voted ignorantly.

The people realized that men were not capable of governing men. What men lack in their ability to govern themselves, they make up for in ingenuity. They made a computer that would govern them. They convened the founding programmers. A group of computer programmers, every gender, every race, every religion and nationality. They convened, and agreed to program the governing computer to govern along three guidelines.

>   maintain order
>   govern without biases
>   govern equally

They then uploaded the combined wisdom of humanity into a mainframe. Books, newspapers, religious text, laws, movies, all were uploaded for the governing computer to pull information, and apply to an algorithm to make laws, to make decisions, and to pass judgement. The founding computer went online. Elected officials ceded their authority to a machine, that they trusted more than themselves and they waited for its first command.

> policemen fired
> can't police effectively
> robot police force

Newly unemployed law enforcement officers were stunned. Factories burst into life, creating robots ordained by the governing computers specifications. The governing computer programmed the robots by the same guidelines he was programed too. The new robot police force hit the streets and all was well until during a traffic stop a police robot shot and killed a driver who was irritated about getting a ticket. Another shot a man for loitering. Police shootings became a common occurrence and the people became fed up... and they revolted.

    brutal overlords
    revolt against the robots
    water cannons work

What went wrong? How did this happen? Was the algorithm wrong? Why would the governing computer police this way? The computer must have queried police and in the material found something that would make it program the robots that way. The founding programmers looked at the past actions of the governing computer. All looked well except...

    search how to police
    newspaper story about
    phlando castille

android
programmed to work like humans
quits job

mechanical gulp
swallowed in metal belly
man-eating robot

unborn child
artificial womb
android wife

screams of mercy
fell on soldered ears
android soldiers

coughs in the hospice
pill dispense robot was there
to see his last breath

warmth of family
why cant I feel this moment?
android downloads loves

feeling emptiness
but you weren't programmed to cry
tears will make you rust

parents sat him down
holding his hand they told him
he was a robot

Dr. West

It's not often that the world is blessed with a great mind, so when we do we cherish it. Well first we slander, and kill them, but when it's all said and done we cherish our great thinkers. Imagine my surprise when I learned of the curious case of Dr. West, whose story I encountered while at the Moorland-Spingarn Research Center. He was one of the celebrated scholars of his day. The masses rushed to hear the sage speak, his books were best sellers, and he was regarded as the conscious of our society. Soon he meddled in the political affairs of the nation, and that is not an acceptable place for a scholar. He meddled like Socrates, and was dealt with thusly. First he was excommunicated from the Ivory Tower, and he drifted quietly into obscurity. The man who had dedicated himself to living the life of the mind was forbade from doing so, and so like Socrates he took Hemlock. Soon his name, and works were forgotten. Until now...

>Schomburg Library
>in ruins rediscovered
>the book "Race Matters"

With a few well directed thoughts I closed my eyes and scanned the database of minds. A mind software that gave me access to all the thoughts, memories, and knowledge of every human being that had uploaded the software to their minds. Upon a person's death the software would upload the persons mind to a searchable archive. People's experiences, knowledge and memories lived on beyond them. Wisdom had become a marketable technology.

> open source thinking
> free access to all your thoughts
> even sexual

Even after a thorough scan I found nothing. This mans name and legacy had been effectively whipped out from history. His story must be told. Who was he? What were his thoughts? What did he believe? The man had been effectively removed from history, where could I find information on him? I went into the antique room and dusted off the computer. Dr. West was from an era when people didn't keep personal diaries. Now historians have discovered a place where those from the past would share their thoughts, and opinions, and at times share their lives in detail. I knew where I could find all the information I needed about Dr. West.

    historic resources
    personal daily accounts
    log on to Facebook

About the Author...

Bio: William Landis is a science fiction poet from North Carolina. He is proud to have been previously published in Scifaikuest, Niteblade, Tales of the Talisman, Star*line, Speculative 66, Parabnormal, Bloodbond, Drabbler, Scryptic, Aphelion, Paper Wasp, Genesis: Anthology of Black Science Fiction, Mindflights, Scierogenous 2, Gatehouse Gazette, Black Petals, Bewildering Stories, and the Dwarf Star Anthology. He is a graduate of North Carolina A&T State University, completing both undergraduate, and graduate work in agriculture. He currently works as an agriculture extension agent where he is currently working on a vermicomposting project, and an Army reserve engineer officer. When he isn't working, he enjoys running, writing, reading, and exploring new places.

William Landis' work also appears in...

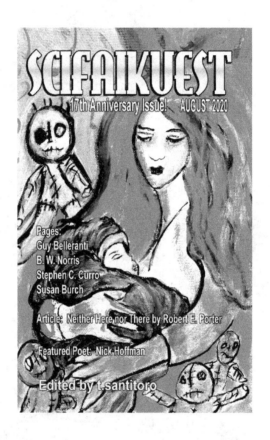

https://www.hiraethsffh.com/product-page/scifaikuest-august-2020

...and in...

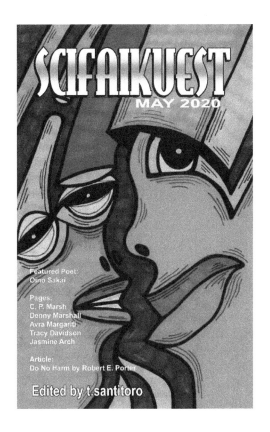

https://www.hiraethsffh.com/product-page/scifaikuest-may-2020

...and in this one he has his own page.

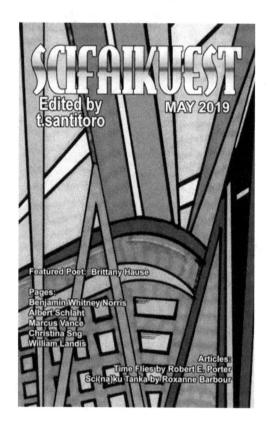

https://www.hiraethsffh.com/product-page/scifaikuest-may-2019

...and here's how he puts his poems together. It's a minimalist poetry handbook...

This handbook contains articles about how to write various minimalist poetry forms such as scifaiku, senryu, sijo, haibun, empat perkataan, ghazals, cinquain, cherita, rengays, rengu, octains, tanka, threesomes, and many more. Each article is written by an expert in that particular poetry form.

https://www.hiraethsffh.com/product-page/minimalism-a-handbook-of-minimalist-genre-poetic-forms

What???
No subscription to
Scifaikuest??

We can fix that . . .

https://www.hiraethsffh.com/product-page/scifaikuest-1

Or get a sample back issue to check us out!

https://www.hiraethsffh.com/shop-1

And a subscription makes a great gift, for a holiday or any time of the year!

CPSIA information can be obtained
at www.ICGtesting.com
Printed in the USA
BVHW032137051022
648810BV00010B/323